U.S. Department
of Transportation
**National Highway
Traffic Safety
Administration**

DOT HS 811 617

April 2012

Tire-Related Factors in the Pre-Crash Phase

DISCLAIMER

This publication is distributed by the U.S. Department of Transportation, National Highway Traffic Safety Administration, in the interest of information exchange. The opinions, findings, and conclusions expressed in this publication are those of the authors and not necessarily those of the Department of Transportation or the National Highway Traffic Safety Administration. The United States Government assumes no liability for its contents or use thereof. If trade names, manufacturers' names, or specific products are mentioned, it is because they are considered essential to the object of the publication and should not be construed as an endorsement. The United States Government does not endorse products or manufacturers.

Suggested APA Citation:

Choi, E-H. (2012, April). Tire-Related Factors in the Pre-Crash Phase. (Report No. DOT HS 811 617). Washington, DC: National Highway Traffic Safety Administration.

1. Report No. DOT HS 811 617	2. Government Accession No.	3. Recipient's Catalog No.
4. Title and Subtitle Tire-Related Factors in the Pre-Crash Phase		5. Report Date April 2012
		6. Performing Organization Code NPO-421
7. Author(s) Eun-Ha Choi, Ph.D. Bowhead Systems Management, Inc. Contractor working at NCSA		8. Performing Organization Report No.
9. Performing Organization Name and Address: Mathematical Analysis Division, National Center for Statistics and Analysis National Highway Traffic Safety Administration 1200 New Jersey Avenue SE., Washington, DC 20590		10. Work Unit No. (TRAIS)
		11. Contract or Grant No.
12. Sponsoring Agency Name and Address Mathematical Analysis Division, National Center for Statistics and Analysis National Highway Traffic Safety Administration 1200 New Jersey Avenue SE., Washington, DC 20590		13. Type of Report and Period Covered NHTSA Technical Report
		14. Sponsoring Agency Code
15. Supplementary Notes		
16. Abstract This study focuses on tire problems as experienced by light vehicles in the pre-crash phase. Of special interest are tire problems such as blowouts or flat tires, tire or wheel deficiency, tire or wheel failure, and tire degradation. The effect of crash factors on a vehicle experiencing such tire problems in the pre-crash phase is examined. Tire tread depth, tire pressure, driving experience, vehicle familiarity, rollover, aggressive driving acts, rollover, vehicle body type, and climatic and road conditions are the candidate factors. The National Motor Vehicle Crash Causation Survey (NMVCCS) data from 2005 to 2007 is used in the statistical analyses. This data, collected at the crash scene, provides information about what happened immediately prior to the crash. A descriptive analysis of this data brings out the differences among the assigned categories of variables in terms of the frequencies of vehicles or tires in each category. The configural frequency analysis confirms the association that certain factors may have with a vehicle experiencing tire problems in the pre-crash phase. Among other findings, the analysis discovered that a vehicle is more likely to experience tire problems when one or more tires are underinflated or the vehicle is running on tires with inadequate tread depth. The emergence of tire problems in the pre-crash phase is significantly more likely than chance if a driver is less familiar with the vehicle or lacks driving experience.		
17. Key Words Tire-related crash vehicles, crash factors, associated factors, critical event, critical reason	18. Distribution Statement Document is available to the public from the National Technical Information Service www.ntis.gov	
19. Security Classif. (of this report) Unclassified	20. Security Classif. (of this page) Unclassified	21. No. of Pages 29	22. Price

TABLE OF CONTENTS

List of Tables iii

List of Figures iv

List of Acronyms v

Executive Summary vi

1. Introduction 1
2. The NMVCCS data 1
3. The NMVCCS perspective of a crash and emergence of tire problems 2
4. Factors associated with tire's crash role 3
5. Analysis methodology 4
6. Analysis results 5
   6.1 Tire damage prior to first harmful event 5
   6.2 Tire pressure 6
   6.3 Tire tread depth 8
   6.4 Vehicle rollover 10
   6.5 Recent experience on this vehicle 11
   6.6 Inexperienced driver 12
   6.7 Roadway related factors 13
   6.8 Aggressive driving acts 14
   6.9 Climatic conditions 14
7. Summary and discussion 15
8. Appendix 17
9. References 19

LIST OF TABLES

Table 1. Categorization of analysis variables 4

Table 2. Observed and expected frequencies by tire's crash role and presence of prior tire damage 6

Table 3. Observed and expected frequencies of tires by tire's crash role and tire inflation status 7

Table 4. Observed and expected frequencies of tires by TPMS use status and tire inflation status 8

Table 5. Observed and expected frequencies of tires by tire's crash role and tire tread depth status 9

Table 6. Observed and expected frequencies by tire's crash role and rollover status 10

Table 7. Distribution of tire-related crash vehicles over rollover status for each vehicle body type 10

Table 8. Observed and expected frequencies by tire's crash role and recent experience to this vehicle 12

Table 9. Observed and expected frequencies by tire's crash role and presence of inexperienced driver ...13

Table 10. Observed and expected frequencies by tire's crash role and presence of roadway-related factors 13

Table 11. Observed and expected frequencies by tire's crash role and presence of driver aggressive acts 14

Table 12. Observed and expected frequencies by tire's crash role and climatic condition 15

LIST OF FIGURES

Figure 1. Three perspectives of tire problems emerging in the pre-crash phase 2

Figure 2. Tire-related crash vehicles and vehicle body type 5

Figure 3. Percentage of tire-related crash vehicles by presence of prior tire damage 6

Figure 4. Percentage of tires of the tire-related crash vehicles in each category of tire inflation status (underinflated, correctly inflated, and overinflated) 7

Figure 5. Percentage of tires of tire-related crash vehicles in each category of tire tread depth 9

Figure 6. Percentages of tire-related crash vehicles for each vehicle body type and rollover status 10

Figure 7. Percentage of tire-related crash vehicles in each category of recent experience to this vehicle ... 11

Figure 8. Percentage of tire-related crash vehicles in each category of inexperienced driver's presence ... 12

Figure 9. Percentage of tire-related crash vehicles in each category of roadway-related factors' presence ... 13

Figure 10. Percentage of tire-related crash vehicles in each category of aggressive driving acts' presence ... 14

Figure 11. Percentage of tire-related crash vehicles by climatic condition 15

LIST OF ACRONYMS

TPMS - tire pressure monitoring systems

NMVCCS - National Motor Vehicle Crash Causation Survey

CFA - configural frequency analysis

SUV - sport utility vehicle

FMVSS - Federal Motor Vehicle Safety Standard

NASS-CDS - National Automotive Sampling System - Crashworthiness Data System

FARS - Fatality Analysis Reporting System

OEM - original equipment manufacturer

EXECUTIVE SUMMARY

This study focuses on tire problems as experienced by light vehicles in the pre-crash phase. Of special interest are these tire-related events: blowouts or flat tires, tire or wheel deficiency, tire or wheel failure, and tire degradation. According to a 2003 NHTSA report, an estimated 414 fatalities, 10,275 non-fatal injuries, and 78,392 crashes occurred annually due to flat tires or blowouts before tire pressure monitoring systems (TPMS) were installed in vehicles.[1] As a result of tire-related safety concerns, NHTSA established two new Federal Motor Vehicle Safety Standards: FMVSS No. 138[2] requires TPMS on all new light vehicles and FMVSS No. 139[3] updated the performance requirements for passenger car and light-truck radial tires. Both of these rules became effective on September 1, 2007. The effects of these rules are expected to continue to increase with time as market penetration increases. This study uses data collected through the National Motor Vehicle Crash Causation Survey in 2005 to 2007 to focus on tire problems experienced by light vehicles in the pre-crash phase. Other factors such as inadequate tread depth, tire underinflation, or extreme climatic conditions could also amplify the emergence of tire problems in this phase. Factors that were analyzed to assess the emergence of tire problems in the pre-crash phase include tire pressure, tread depth, vehicle body type, vehicle rollover, driver's familiarity with the vehicle, driving experience, aggressive driving behavior, roadway-related factors, and climatic conditions.

The NMVCCS recorded the sequence of events occurring in the pre-crash phase, including those events related to tires. In the survey, tire problems experienced by vehicles in the pre-crash phase are recorded as associated factors, critical pre-crash events, or critical reasons, denying any implication that these problems actually caused the crash. This study uses the NMVCCS data in a descriptive analysis to highlight the differences among categories of the associated factors. Configural frequency analysis is conducted to study the association of these factors with the emergence of tire problems.

The NMVCCS data is a sample of 5,470 crashes representing 2,188,970 crashes at the national level. In 9 percent of these crashes, one or more vehicles experienced tire problems in the pre-crash phase. Correspondingly, of the estimated 3,889,770 vehicles involved in the NMVVCS crashes, 5 percent experienced tire problems in the pre-crash phase. Fifty percent of the tire-related crashes were single-vehicle crashes while only 31 percent of crashes where tire-related crash factors were not cited were single-vehicle crashes. Some of the results from this study are listed below.

- Of the tires that were underinflated by more than 25 percent of the recommended pressure, approximately 10 percent were in vehicles that experienced tire problems in the pre-crash phase. In contrast, among the correctly inflated tires, a much smaller percentage (3.4%) belongs to vehicles that experienced tire problems. Thus, underinflation is not the only cause of tire problems; however, when tires are underinflated by 25 percent or more, tires are 3 times as likely to be cited as critical events in the pre-crash phase.

- With at least one or more tires with lower tread depths (between 0 and 4/32"), vehicles experienced tire problems during crash occurrence significantly more than chance. Of tires with tread depth in the range 0 to 2/32", about 26 percent were in vehicles that experienced tire problems in the pre-crash phase while only 8 percent of tires with tread depth in the range 3/32 to 4/32" were in such vehicles.

- The percentage of vehicles experiencing tire problems is significantly higher among vehicles that rolled over as compared to vehicles that did not roll over for all vehicle body types: passenger cars, pickups, SUVs, and vans. Of all SUVs experiencing tire problems in the pre-crash phase, 45 percent rolled over. For the other body types (passenger cars, pickups, and vans), fewer than 25 percent of the vehicles experiencing tire problems rolled over. Thus, tire problems experienced in the pre-crash phase were more likely to result in a rollover in SUVs than in other vehicle types.

- When drivers were less familiar with the vehicles they were driving, the vehicles experienced tire problems in the pre-crash phase significantly more than chance. This was also the case when drivers were inexperienced and lacked sufficient driver training. Thus, it is likely that inexperienced drivers and drivers not familiar with the vehicles they are driving pay less attention to tires and tire pressure.

- A significantly higher percentage (11.2%) of vehicles were observed to experience tire problems when one or more roadway-related factors (e.g., wet road, road under water, slick surface) were present in the pre-crash phase as compared to when no roadway-related factors were cited (3.9%). Thus, the vehicles running under adverse roadway conditions may become more vulnerable to tire problems.

1. INTRODUCTION

In order for a vehicle to handle safely and to use fuel economically, the vehicle's tires should be in good condition. Good condition requires regular monitoring and timely maintenance of all tires on, or associated with, the vehicle. Nevertheless, it is not uncommon to find vehicles on the road, running on one or more underinflated/overinflated tires or tires with inadequate tread depth. Tire pressure below the recommended pressure can cause high heat generation that in turn can cause rapid tire wear and blowout. Similarly, inadequate tread depth can also cause blowouts of tires. Tire-related events such as tire failure or blowout resulting from tire deficiencies or other factors are risky and often add to the likelihood of crash occurrence. According to a 2003 NHTSA report, an estimated 414 fatalities, 10,275 non-fatal injuries, and 78,392 crashes occurred annually due to flat tires or blowouts before tire pressure monitoring systems were installed in vehicles.[1]

When a vehicle starts experiencing tire problems in the pre-crash phase, i.e., immediately prior to the collision, the time window for attempting a crash avoidance maneuver is extremely small. This makes the vehicle vulnerable to crash involvement. Also, the risk of collision may be enhanced if one or more crash factors are present in this phase of the crash. For example, during crash occurrence, tire blowout of a vehicle running on a wet road or driven by an inexperienced driver may make the crash unavoidable. In one of the investigated crashes, two rear tires of a crash-involved pickup truck had only 1/32 inch of tread depth.[4] The driver felt the rear-end of this vehicle "slip" during crash occurrence, probably due to tire failure. This happened when it was raining and the road was wet. To reduce the number of crashes that are attributable to tire problems, it is important to study the crash-involved vehicles that experienced tire problems in the pre-crash phase. The knowledge about the effect of other crash factors on a vehicle experiencing tire problems in the pre-crash phase can provide a better perspective of the crashes that may be attributed to tire problems.

The choice of information available for this purpose is limited. The Indiana Tri-Level Study[5] data, collected in 1979, has tire information but is outdated. Since then, much has changed – the use of radial tires on vehicles has increased and so has the availability of tire pressure monitoring devices such as TPMS. Some of the other databases that contain the tire-related information are National Automotive Sampling System - Crashworthiness Data System and the Fatality Analysis Reporting System. Even though the data pertaining to crash-involved vehicles is updated annually, the information is compiled much later after the crash has occurred. Thus, these databases provide little clue on what and how tire problems were experienced by vehicles during crash occurrence. Additionally, both databases lack information on tire pressure. To obtain the firsthand information about several aspects of crashes, NMVCCS was conducted by NHTSA's National Center for Statistics and Analysis in 2005 to 2007. This includes information about tire-related events such as tire failure and tire blowout that occurred in the pre-crash phase, as well as other factors present in the crash.

2. THE NMVCCS DATA

During the 3-year period January 2005 to December 2007, NMVCCS collected driver-, vehicle-, roadway-, and environment-related information from 6,949 crashes. Each of these crashes occurred between 6 a.m. and midnight and resulted in a harmful event associated with a vehicle in transport. Additionally, at least one of the first three vehicles in these crashes was a light passenger vehicle towed due to damage. The aim of NMVCCS was to record an account of the sequence of events that led to the crash. To achieve this, the crashes were investigated immediately after the crash occurrence without assigning the fault to the driver, vehicle, or environment.

The NMVCCS data has certain limitations as in any survey. The small sample sizes, due to a large number of unknowns or data segmentation required for certain types of analyses, may affect the precision of the estimates. The information in this survey was recorded from driver and witness interviews, vehicle assessment, and evaluation of the roadway infrastructure. Therefore, caution is needed when interpreting the results, as some of the variables used in the analysis are subjective in nature. The NMVCCS data also contains multiple-choice variables whose attributes may define overlapping categories. This may violate the assumption of mutual exclusiveness required for certain types of analyses.

Of the total 6,949 crashes investigated during July 2005 to December 2007, sampling weights were assigned to 5,470 crashes to yield a nationally representative sample. The present study analyzes the data pertaining to these weighted crashes.

3. THE NMVCCS PERSPECTIVE OF A CRASH AND EMERGENCE OF TIRE PROBLEMS

A crash in NMVCCS is considered as a simplified linear chain[6] of events comprised of "crash-associated factors," "movement prior to critical crash envelope," "critical pre-crash event," and "critical reason for the critical pre-crash event" (Figure 1). Among these elements, the critical pre-crash event documents the circumstances that made the crash imminent. The movement prior to critical crash envelope refers to movement of the vehicle immediately before the critical pre-crash event. The crash-associated factors document factors that might have played a role in crash occurrence. The critical reason is the immediate reason for the critical event and is often the last failure in the causal chain (i.e., closest in time to the critical pre-crash event).

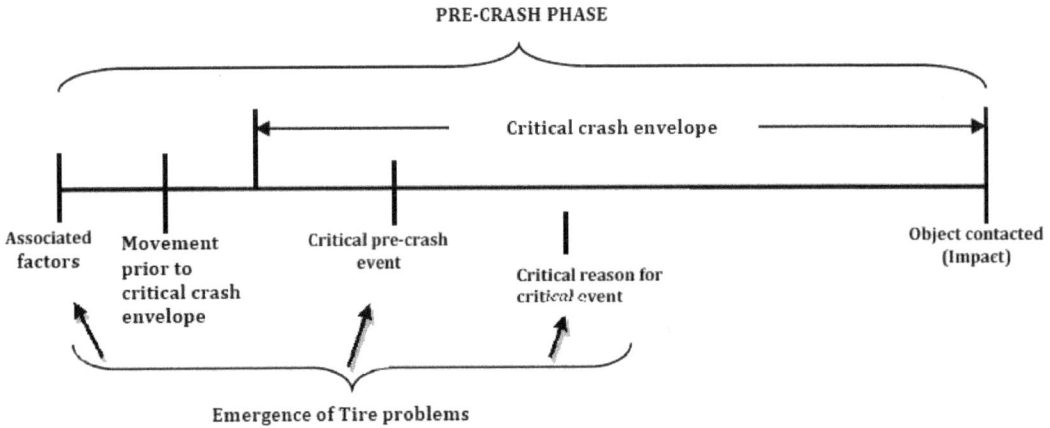

Figure 1. Three perspectives of tire problems emerging in the pre-crash phase

This study considers tire problems in the above perspective, i.e., the emergence of tire problems in the pre-crash phase as an associated factor, the critical pre-crash event, or the critical reason for the critical event. However, none of these implies that a particular tire problem caused the crash. In the subsequent discussion, these tire problems in the pre-crash phase are referred to as "tire-related crash factors" and are defined as follows.

Tire/wheel deficiency – assessed as an associated factor –
the variable that indicates if the vehicle experienced a tire deficiency/malfunction (e.g., blowout, air out, etc.) in the pre-crash phase.

Blowout or flat tires – assessed as the critical pre-crash event –
the variable that shows if blow out or flat tire caused loss of vehicle control when in motion.

Tires/wheels failed, other tire degradation – assessed as the critical reason –
the variable that records catastrophic failures (e.g., blowouts, tread separations, wheel separations) and tire degradation (e.g., bald and/or underinflated tires) that may degrade the vehicle's handling characteristics.

The focus of this study is on the vehicles that experienced at least one of these tire problems in the pre-crash phase, namely tire/wheel deficiency, blowout or flat tire, and tires or wheels failed or other tire degradation.

For the subsequent analysis and discussion, the composite variable "tire's crash role" is defined as

$$\text{Tire's crash role} = \begin{cases} \text{Active, if at least one of the tire-related crash factors is present in the pre-crash} \\ \text{phase for the vehicle,} \\ \text{Inactive, if none of the tire-related crash factors is present in the pre-crash phase} \\ \text{for the vehicle.} \end{cases}$$

A crash-involved vehicle for which the "tire's crash role" is active is referred to as a "tire-related crash vehicle" and a crash in which such a vehicle is involved, a "tire-related crash." For comparison purposes, the study also considers vehicles for which the tire's crash role is inactive. These vehicles are referred to as "other crash vehicles."

4. FACTORS ASSOCIATED WITH TIRE'S CRASH ROLE

The tire's crash role may be associated with other driver-, vehicle-, or environment-related factors. To study the association of the tire's crash role with other crash factors, the following variables are considered.

Vehicle factors
- *Tire pressure:* tire inflation status
- *Tire tread depth:* tire tread depth measured to the nearest 1/32nd of an inch
- *Tire damage prior to first harmful event:* the pre-crash flaws or damage in each tire of the vehicle (e.g., complete tread separation, partial tread separation)
- *TPMS*: if tire pressure monitoring system is in use or not
- *Vehicle body type:* a vehicle body type describing the general configuration/shape and distinguishing characteristics of the motor vehicle
- *Rollover:* whether or not the vehicle rolled over

Driver factors
- *Inexperienced driver:* the presence of a driver with a lack of training or driving experience
- *Recent experience driving this vehicle*: the driver familiarity with this vehicle in terms of the number of times the driver drove this vehicle in the past three months
- *Aggressive driving acts:* the presence of one or more of the aggressive driving acts such as speeding, tailgating, rapid/frequent lane changes/weaving, accelerating rapidly from stop, and stopping suddenly.

Environmental factors
- *Road related factor*: the presence of one or more roadway-related factors such as wet roads, road under water, slick surface, road washed out, potholes, deteriorated road edges, etc.
- *Month of the crash:* month of the year in which crash occurred

For the purpose of analyses, these variables are categorized as shown in Table 1.

Table 1. Categorization of analysis variables

Variable	Categories used in analysis
Tire's crash role	Active, Inactive
Improper tire pressure (in percent)	Less than -25, -25 to -10, -10 to 0, 0, 0 to 10, 10 to 25, Greater than 25
Tire tread depth (in 1/32nd of an inch)	0-2, 3-4, 5-6, 7+
Tire damage prior to first harmful event	Yes, No
TPMS in use	Yes, No
Vehicle body type	Passenger cars, Pickups, SUVs, Vans
Rollover	Yes, No
Inexperienced driver	Yes, No
Recent experience driving this vehicle (in last three months)	1-10 times, More than 10 times
Aggressive driving act	One or more present, None
Roadway related factor	One or more present, None
Climatic condition	Cold (Nov. to Feb.), Hot (July to Sept.), Mild (Mar. to June, Oct.)

5. ANALYSIS METHODOLOGY

Descriptive and configural frequency analyses (CFA) are conducted to study tire problems as experienced by vehicles in the pre-crash phase. The purpose of the descriptive analysis is to bring out differences among the vehicle categories as defined by the attributes of the variables, specified in Table 1. This is done in terms of the percent frequency of vehicles falling in each of these categories. CFA[7, 8] is conducted to identify other crash factors that are associated with the emergence of tire problems. In certain attributes of a factor, a vehicle is more likely to experience tire problems as compared to others. For instance, a tire with inadequate (less than 2/32") tread depth may make the vehicle more vulnerable to experiencing tire problems as compared to the tire that has adequate tread depth of 5/32" or more. Identifying such profiles that may have an effect on tire's crash role will be useful in developing and implementing crash prevention measures.

CFA and its interpretation:
CFA is a multivariate statistical technique that searches patterns of variables' categories, which occurred more often or less often than expected under chance alone. The population units (crash-involved vehicles or tires, in the present case) are segmented by cross-tabulation of the variables. Each combination of the variables' categories used in the segmentation, referred to as a configuration, characterizes a profile of the population unit (e.g., a vehicle experienced tire problems in the pre-crash phase while being driven on a wet road). By comparing the observed and expected frequencies for each configuration, CFA explores the association of tire's crash role with other factors of interest such as driving experience, tire pressure, tread depth, aggressive driving acts, climatic conditions, etc. Since the expected frequency in this technique is estimated based on the assumption of no association between factors, the expected frequency is presumed to be the outcome of chance alone. Thus, a statistically significant difference between the observed and expected frequency for a configuration provides evidence of the association a factor has with a vehicle experiencing tire problems in the pre-crash phase. The Z-statistic is used to confirm if the difference (positive or negative) is statistically significant. Computational details of the Z-statistic are provided in the Appendix. All inferences made through CFA bear a 99 percent confidence level. This is the Bonferroni adjusted level of the initially set 95 percent confidence level. In the subsequent discussion, for the sake of brevity, the expression "a vehicle experiencing tire problems" is used for the expression "a

vehicle experiencing tire problems in the pre-crash phase." Such a vehicle is also referred to as a "tire-related crash vehicle."

The results of descriptive analysis are presented as bar charts that show percent frequencies of vehicles falling in different categories of a variable, specified in Table 1. CFA results are presented in tables that show configurations of variables (defining vehicle profiles) and the corresponding observed and expected frequencies, as well as the Z-values. The statistical software SAS 9.1.3 is used for these analyses.[9]

6. ANALYSIS RESULTS

The 5,470 NMVCCS crashes represent an estimated 2,188,970 crashes at the national level. Approximately 9 percent (189,917) of the estimated total were "tire-related crashes." In terms of vehicles, of the estimated 3,894,507 vehicles involved in the NMVCCS crashes, 197,421 (about 5% of the estimated total) are "tire-related crash vehicles." About 50 percent of the tire-related crashes were single-vehicle crashes while only about 31 percent of other crashes were single-vehicle crashes.

Figure 2(a) shows the percentages of tire-related crash vehicles in each vehicle body type, shown as shaded portions of bars and of "other crash variables" as un-shaded portions. About 6 percent of passenger cars, 4.6 percent of SUVs, 4.3 percent of pickups, and 3.5 percent of vans are tire-related crash vehicles. Figure 2(b) displays the percentage distribution of the tire-related crash vehicles over vehicle body types. Among the tire-related crash vehicles, the light passenger vehicles account for the largest percentage (66.3%), followed by SUVs (17.4%), pickups (11.1%), and vans (4.9%).

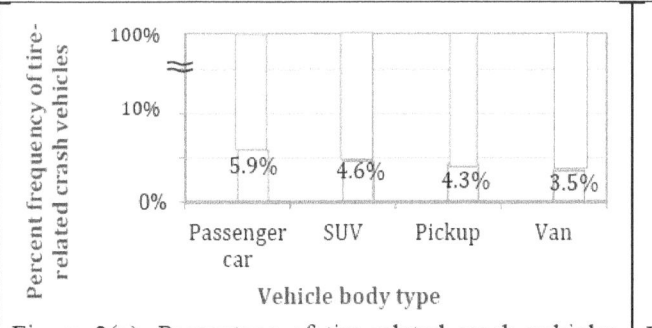

Figure 2(a). Percentage of tire-related crash vehicles in each vehicle body type

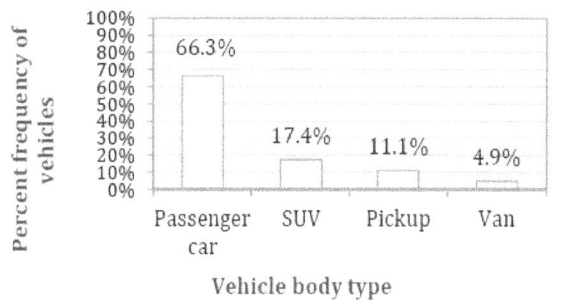

Figure 2(b). Percentage distribution of tire-related crash vehicles over vehicle body type

Figure 2. Tire-related crash vehicles and vehicle body type
(Data Source: NMVCCS 2005-2007)

6.1 TIRE DAMAGE PRIOR TO FIRST HARMFUL EVENT

The variable "tire damage prior to first harmful event" records the flaws or damage in a tire, such as complete or partial tread separation prior to the first harmful event (i.e., the first event that caused a fatal or nonfatal injury or property damage). The NMVCCS data shows that of all the vehicles that had prior tire damage to one or more of their tires, 31.6 percent experienced tire problems and about 68.4 percent did not experience tire problems (Figure 3). On the other hand, among vehicles with no prior tire damage, only 4.5 percent were tire-related crash vehicles and 95.5 percent were other crash vehicles. Table 2 presents the results of CFA to detect the association of prior tire damage with tire's crash role.

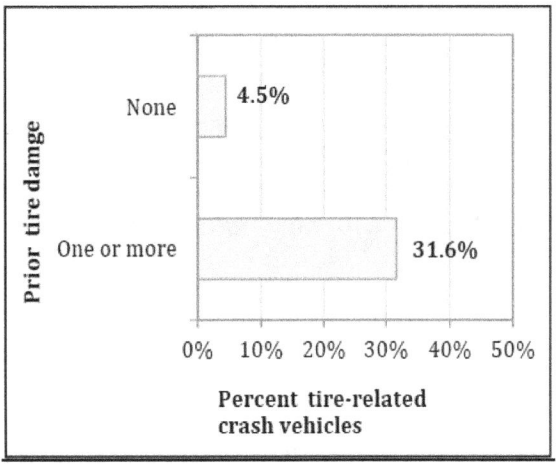

Figure 3. Percentage of tire-related crash vehicles by presence of prior tire damage
(Data Source: NMVCCS 2005-2007)

The results show that when there was prior tire damage, significantly more than expected vehicles were observed to experience tire problems under the assumption that tire's crash role has no association with prior tire damage (Z-value = 11.7). The negative Z-value, -1.8 shows that when there was no prior tire damage, vehicles were observed to experience tire problems in the pre-crash phase significantly less than chance.

Table 2. Observed and expected frequencies by tire's crash role and presence of prior tire damage

Tire's crash role (Vehicle classification)	Prior tire damage	Observed	Expected	Z-value
Active (Tire-related crash vehicle)	One or more	31,234	5,173	11.7*
	None	152,540	178,601	-1.8*
Inactive (Other crash vehicle)	One or more	67,603	93,664	-2.6*
	None	3,259,734	3,233,673	0.4

*Statistically significant at 95 percent confidence level
(Data Source: NMVCCS 2005-2007)

6.2 TIRE PRESSURE

NMVCCS records both the recommended and measured tire pressures of each tire of the crash-involved vehicles. These measurements can be used to determine a tire's inflation status as:

- Underinflated, if the recommended tire pressure exceeds the measured pressure
- Overinflated, if the measured tire pressure exceeds the recommended pressure
- Correctly inflated, if the measured tire pressure is the same as the recommended pressure

In addition to correct tire inflation, three levels of underinflation and overinflation are considered in the analysis. These include 0 to 10, 10 to 25, and greater than 25 percent underinflation or overinflation compared to the recommended pressure. TPMS is required to provide a warning to the driver when one or more tires are 25 percent or more below the recommended pressure on the tire placard.

Figure 4 shows that of all the tires underinflated by more than 25 percent of the recommended pressure, about 10 percent belonged to tire-related crash vehicles, which is the highest among the three categories of underinflated tires. About 4 percent of the tires that were underinflated by less than 10 percent and about 6 percent of those underinflated by 10 to 25 percent belonged to tire-related crash vehicles. Similarly, of all the tires overinflated by more than 25 percent, about 7 percent were mounted on tire-related crash vehicles. Figure 4 shows increasing percentages of underinflated or overinflated tires belonging to tire-related crash vehicles with increasing levels of underinflation or overinflation. Only 3.4 percent of the correctly inflated tires belonged to tire-related crash vehicles. The percentage of other crash vehicles for each category of tire inflation status can be obtained by subtracting the percentage of tire-related crash vehicles, presented in Figure 4, from 100. For example, of the correctly inflated tires, 3.4 percent belonged to tire-related crash vehicles and 96.6 percent belonged to other crash vehicles. Note that these profiles do not include cases that are recorded as "no original equipment manufacturer (OEM) wheel at this location," "flat tire" or "unknown."

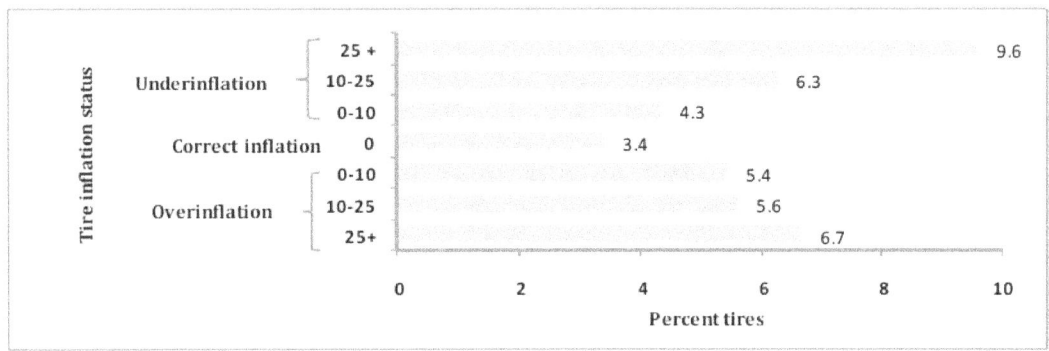

Figure 4. Percentage of tires of the tire-related crash vehicles in each category of tire inflation status (underinflated, correctly inflated, and overinflated)
(Data Source: NMVCCS 2005-2007)

The effect of tire inflation on tire's crash role during the pre-crash phase is studied by conducting CFA. The results are presented in Table 3. The positive Z-value 5.1 shows that vehicles with tires underinflated by more than 25 percent of the recommended pressure experienced tire problems significantly more than expected under the assumption that tire's crash role has no association with tire inflation. The negative Z-value -3.3 suggests that with correctly inflated tires, significantly less than expected vehicles were observed to experience tire problems in the pre-crash phase.

Table 3. Observed and expected frequencies of tires by tire's crash role and tire inflation status

Tire's crash role (Vehicle classification)	Tire inflation status (percent of recommended pressure)		Observed	Expected	Z-value
Active (Tire-related crash vehicle)	Underinflation	25+	103,735	62,042	5.1*
		10-25	127,183	116,600	1.0
		0-10	81,806	108,334	-1.7
	Correct tire pressure	0	36,439	61,714	-3.3*
	Overinflation	0-10	78,447	82,976	-0.5
		10-25	75,037	76,942	-0.2
		25 +	43,181	37,220	1.0
Inactive (Other crash vehicle)	Underinflation	25+	975,631	1,017,324	-1.2
		10-25	1,901,357	1,911,939	-0.2
		0-10	1,802,934	1,776,407	0.5
	Correct tire pressure	0	1,037,222	1,011,947	0.7
	Overinflation	0-10	1,365,123	1,360,594	0.1
		10-25	1,263,560	1,261,655	0.0
		25 +	604,352	610,313	-0.2

*Statistically significant at 95 percent confidence level
(Data Source: NMVCCS 2005-2007)

One of the devices used for monitoring tire inflation is the TPMS. The effect of TPMS use in maintaining tire pressure is studied using CFA. Note that the category "TPMS not in use" includes vehicles not equipped with TPMS as well as vehicles equipped with TPMS that was not in use. TPMS was not required on all newly manufactured light vehicles until September 1, 2007. Therefore, the NMVCCS data from 2005-2007 was not able to capture a large number of vehicles equipped with TPMS. Only two percent of the vehicles included in NMVCCS data were equipped with TPMS.

The results are shown in Table 4. The negative Z-value -2.0 shows that significantly less than expected tires were observed to be extremely underinflated (more than 25%of the recommended pressure) when TPMS was used, where expected frequencies were obtained under the assumption that TPMS use status has no association with tire inflation status. The positive Z-value 2.8 suggests that when TPMS was used, correctly inflated tires were significantly more likely than chance. Two Z-values -6.4 show that significantly less than expected tires were overinflated by more than 10 percent of the recommended pressures when this device was used.

Table 4. Observed and expected frequencies of tires by TPMS use status and tire inflation status

TPMS in use	Extent of underinflation or overinflation		Observed	Expected	Z-value
Yes	Underinflation	25 +	7,719	11,328	-2.0*
		10-25	30,846	23,611	1.6
		0-10	29,331	25,538	0.7
	Correct tire pressure	0	43,841	19,040	2.8*
	Overinflation	0-10	34,337	34,024	0.1
		10-25	19,115	37,617	-6.4*
		25 +	6,960	20,991	-6.4*
No	Underinflation	25 +	450,026	446,417	0.2
		10-25	923,278	930,513	-0.3
		0-10	1,002,639	1,006,433	-0.1
	Correct tire pressure	0	725,572	750,372	-0.9
	Overinflation	0-10	1,340,574	1,340,887	0.0
		10-25	1,500,967	1,482,466	0.5
		25 +	841,291	827,259	0.5

*Statistically significant at 95 percent confidence level
(Data Source: NMVCCS 2005-2007)

FMVSS No. 138 requires TPMS on all light vehicles manufactured after September 1, 2007, that will be sold in the U.S.[2] Specifically, the TPMS must alert drivers when the inflation pressure in one or more of their tires falls below 75 percent of the vehicle manufacturer's recommended cold inflation pressure.

6.3 TIRE TREAD DEPTH

An adequate tire tread depth on all tires of a vehicle is important to maintain proper grip on the road under different road conditions. NHTSA recommends that tires should be replaced when the tread depth is 2/32". As a result, FMVSS No. 139 – *New pneumatic radial tires for light vehicles,* requires treadwear indicators that enable a person, through visual inspection, to determine if the tire tread depth is at least one sixteenth of an inch (or 2/32").[3] NMVCCS records tire tread depth to the nearest 1/32 of an inch measured on the shallowest grove of the tread. Four ranges of tread depths, namely 0-2/32", 3- 4/32", 5-6/32", and above 7/32" are considered in the analysis. The data show that of all the tires observed with tread depth between 0 and 2/32", 26.2 percent were mounted on tire-related crash vehicles (Figure 5). In regard to tires with adequate tire tread depth, the data show that 8 percent of the tires with tread depth in the range 3-4/32", 4 percent in the range 5-6/32", and 2.4 percent above 7/32" belonged to tire-related crash vehicles. With tire tread depth in the range 0-2/32", vehicles were observed to experience tire problems in the pre-crash phase 3 times more than vehicles with tread depth in the range 3-4/32". The percentage of other crash vehicles for each category of tire tread depth can be obtained by subtracting the percentage of tire-related crash vehicles, presented in Figure 5, from 100. For example, of the tires with tread depth in the range 3-4/32", 26.2 percent belonged to tire-related crash vehicles and 73.8 percent other crash vehicles.

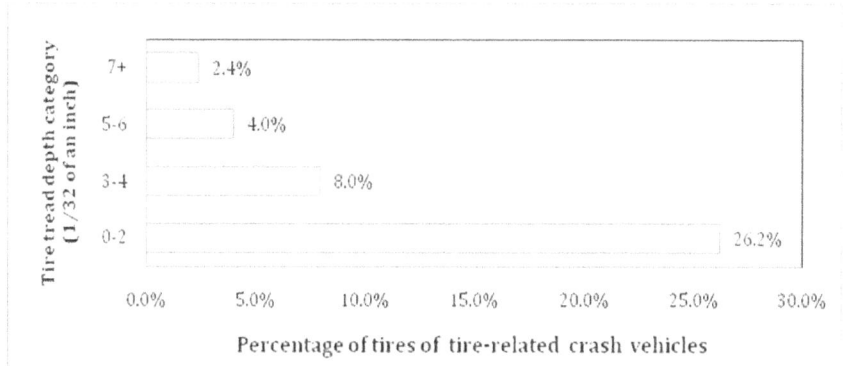

Figure 5. Percentage of tires of tire-related crash vehicles in each category of tire tread depth
(Data Source: NMVCCS 2005-2007)

CFA was conducted to study association of tread depth with tire's crash role. The results in Table 5 show that significantly more than expected vehicles experienced tire problems when they were running on tires with tread depth between 0 and 4/32" (Z-values 24.8 and 3.9), where expected frequencies were obtained under the assumption that tire's crash role has no association with tire tread depth. With tread depth above 4/32", vehicles experiencing tire problems were significantly less likely than chance (Z-values -5.2 and -10).

Table 5. Observed and expected frequencies of tires by tire's crash role and tire tread depth status

Tire's crash role (Vehicle classification)	Tire tread depth (1/32")	Observed	Expected	Z-value
Active (Tire-related crash vehicle)	0-2	262,869	58,593	24.8*
	3-4	188,763	138,570	3.9*
	5-6	216,085	315,050	-5.2*
	7+	108,080	263,585	-10.0*
Inactive (Other crash vehicle)	0-2	740,085	944,361	-5.5*
	3-4	2,183,177	2,233,370	-0.9
	5-6	5,176,719	5,077,754	1.2*
	7+	4,403,783	4,248,278	2.1*

*Statistically significant at 95 percent confidence level
(Data Source: NMVCCS 2005-2007)

6.4 VEHICLE ROLLOVER

Rollovers are mostly single-vehicle crashes because they usually do not involve a collision with another vehicle in transport. Some of the possible reasons for a rollover are tire blowout, loss of tire tread, tire belt peel off, tread separation, and tire bead unseating. These tire-related events can make a vehicle lose control, especially at high speeds, and eventually rollover.

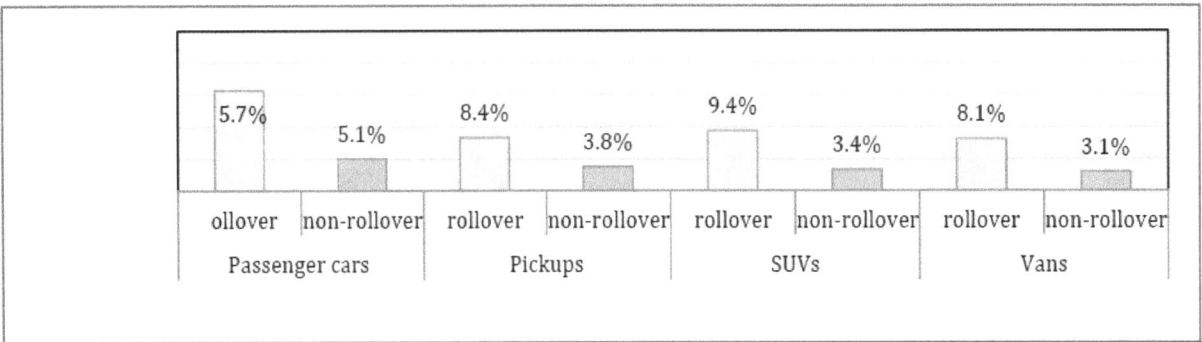

Figure 6. Percentages of tire-related crash vehicles for each vehicle body type and rollover status
(Data Source: NMVCCS 2005-2007)

Table 6. Observed and expected frequencies by tire's crash role and rollover status

crash role classification)	Rollover	Observed	Expected	Z-value
Active (Tire-related crash vehicle)	Yes	51,687	23,130	5.2*
	No	145,733	174,290	-2.1*
Inactive (Other crash vehicle)	Yes	404,602	433,159	-1.0*
	No	3,292,484	3,263,926	0.4*

*Statistically significant at 95 percent confidence level
(Data Source: NMVCCS 2005-2007)

Table 7. Distribution of tire-related crash vehicles over rollover status for each vehicle body type

Vehicle body type	Rollover status	Estimated number of tire-related crash vehicles	Percent distribution of tire-related crash vehicles by rollover status in each vehicle body type
Passenger cars	Rollover	28,948	22.1%
	Non-rollover	101,988	77.9%
Pickups	Rollover	5,323	24.3%
	Non-rollover	16,604	75.7%
SUVs	Rollover	15,386	44.9%
	Non-rollover	18,863	55.1%
Vans	Rollover	2,031	21.0%
	Non-rollover	7,645	79.0%
Total	Rollover	51,687	26.3%
	Non-rollover	145,100	73.7%

(Data Source: NMVCCS 2005-2007)

Figure 6 shows the percentages of tire-related crash vehicles among rollover and non-rollover vehicles by vehicle body types. Of rolled-over passenger cars, 15.7 percent were tire-related crash vehicles, which is the highest as compared to other body types: SUVs (9.4%), pickups (8.4%) and vans (8.1%). Among non-rollover passenger cars, 5.1 percent were tire-related crash vehicles. For all vehicle body types, the percentage of tire-related crash vehicles among rolled-over vehicles is much higher than the percentage among non-rollover vehicles. The percentage of other crash vehicles for each vehicle body type can be obtained by subtracting the percentage of tire-related crash vehicles, presented in Figure 6, from 100. For example, among rollover passenger cars, 15.7 percent were tire-related crash vehicles and 84.3 percent were other crash vehicles.

Analysis is conducted to see if rollover is associated with tire's crash role. CFA results in Table 6 show that significantly more than expected vehicles experienced tire problems in the event of rollover, as indicated by the positive Z-value 5.2, where expected frequencies were obtained under the assumption that tire's crash role has no association with rollover. Vehicles experiencing tire problems were less likely than chance if they do not rollover (Z-value -2.1).

When only tire-related crash vehicles are considered (see Table 7), the highest percentage of rollover is among SUVs (44.9%) as compared to other body types, passenger cars, pickups, and vans (less than 25%). In contrast, the lowest percentage of non-rollover was among SUVs (55.1%).

6.5 RECENT EXPERIENCE ON THIS VEHICLE

A driver's ability to handle a vehicle safely depends to a certain extent on his/her familiarity with the vehicle. NMVCCS recorded this information as driver's experience on the vehicle he/she was driving. This is expressed as the number of times the driver drove this vehicle in the past three months. Figure 7 shows that of all the crash-involved vehicles that were driven 1 to 10 times by their drivers, 8 percent experienced tire problems. This percentage is significantly lower among vehicles that had been driven more than 10 times by their drivers (5.2%). The percentage of other crash vehicles (92%) among all the crash-involved vehicles that were driven 1 to 10 times by their drivers can be obtained by subtracting the percentage of tire-related crash vehicles (8%), presented in Figure 7, from 100. Similarly, the percentage of other crash vehicles among the crash vehicles that were driven more than 10 times is obtained as 94.8 percent.

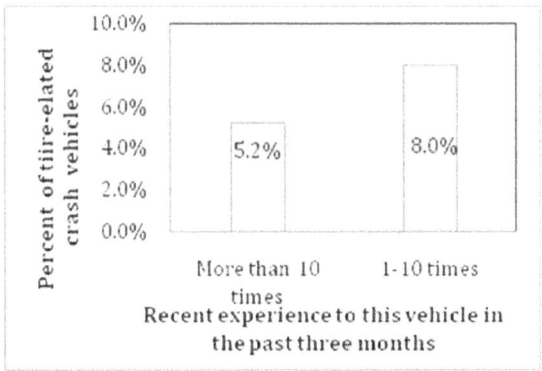

Figure 7. Percentage of tire-related crash vehicles in each category of recent experience to this vehicle.
(Data Source: NMVCCS 2005-2007)

CFA analysis is conducted to see if tire's crash role is associated with driver's experience on the subject vehicle. The Z-value -3.6 in Table 8 suggests that a vehicle driven by a person who is familiar with the vehicle is significantly less likely to experience tire problems than expected under the assumption that tire's crash role has no association with driver's experience on the subject vehicle.

Table 8. Observed and expected frequencies by tire's crash role and recent experience to this vehicle

Tire's crash role (Vehicle classification)	Number of times driving this vehicle in the past three months	Observed	Expected	Z-VALUE
Active (Tire-related crash vehicle)	1 to 10 times (Less familiar)	16,138	14,669	0.3
	More than 10 times (More familiar)	139,975	194,371	-3.6*
Inactive (Other crash vehicle)	1 to 10 times (Less familiar)	186,292	187,761	-0.1
	More than 10 times (More familiar)	2,542,332	2,487,937	1.0*

*Statistically significant at 95 percent confidence level
(Data Source: NMVCCS 2005-2007)

6.6 INEXPERIENCED DRIVER

In addition to driver's experience to the subject vehicle, driving experience in general plays a significant role in handling a vehicle safely. In addition, an inexperienced driver may not know how to properly maintain the vehicle, including tires or they may pay less attention to tires or tire pressure. NMVCCS records a driver as "inexperienced" if the driver lacks training or has less than a year's driving experience.

Figure 8 shows that of all the vehicles observed with inexperienced drivers, about 12 percent were tire-related crash vehicles. On the other hand, among vehicles driven by experienced drivers, tire-related crash vehicles accounted for a significantly smaller percentage (4.8%).

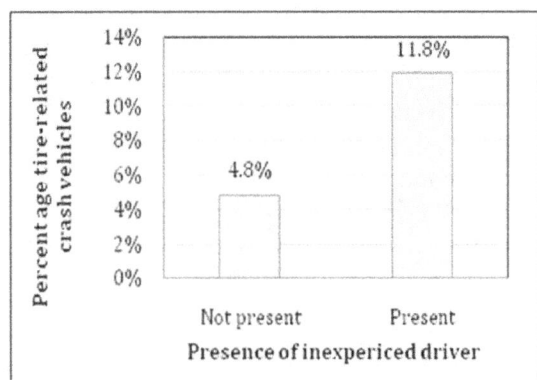

Figure 8. Percentage of tire-related crash vehicles in each category of inexperienced driver's presence
(Data Source: NMVCCS 2005-2007)

The percentage of other crash vehicles for each category of inexperienced driver's presence can be obtained by subtracting the percentage of tire-related crash vehicles from 100. For example, of all crash-involved vehicles observed with inexperienced drivers, about 88 percent were other crash vehicles while about 12 percent were tire-related crash vehicles.

CFA results in Table 9 show that significantly more than expected vehicles experiencing tire problems during crash occurrence were observed with inexperienced drivers (Z-value = 3.7), where expected frequencies were obtained under the assumption that tire's crash role has no association with a driver's driving experience.

Table 9. Observed and expected frequencies by tire's crash role and presence of inexperienced driver

Tire's crash role (Vehicle classification)	Inexperienced driver	Observed	Expected	Z-value
Active (Tire-related crash vehicle)	Yes	28,439	12,981	3.7*
	No	133,380	148,838	-1.2*
Inactive (Other crash vehicle)	Yes	211,622	227,080	-0.9*
	No	2,619,015	2,603,557	0.3

*Statistically significant at 95 percent confidence level
(Data Source: NMVCCS 2005-2007)

6.7 ROADWAY RELATED FACTORS

The NMVCCS data shows that when road-related factors (wet road, road under water, slick surface, or road washed out) were present during crash occurrence, 11.2 percent of the crash-involved vehicles were tire-related crash vehicles (Figure 9). In contrast, in the absence of these factors, only 3.9 percent of vehicles were as such. The percentage of other crash vehicles for each category of roadway-related factors' presence can be obtained by subtracting the percentage of tire-related crash vehicles from 100. For example, when road-related factors were present, about 89 percent were other crash vehicles while about 11 percent were tire-related crash vehicles.

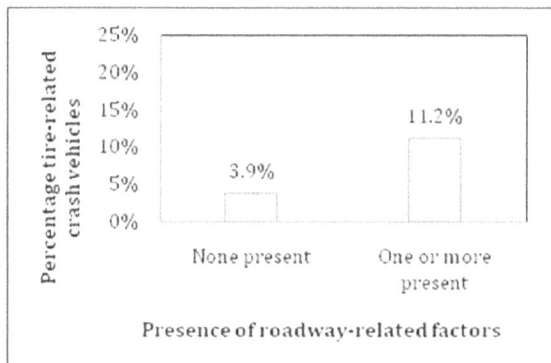

Figure 9. Percentage of tire-related crash vehicles in each category of roadway-related factors' presence.
(Data Source: NMVCCS 2005-2007)

CFA results in Table 10 shed light on the effect of adverse road conditions on tire's crash role. Z-value 6.9 shows that significantly more than expected vehicles experienced tire problems under adverse roadway conditions. When none of such conditions was present, significantly less than expected vehicles experienced tire problems (Z-value -2.6), where the expected frequencies were obtained under the assumption that tire's crash role has no association with presence of roadway-related factors. The results of this analysis show that vehicles running under adverse roadway conditions may become more vulnerable to tire problems.

Table 10. Observed and expected frequencies by tire's crash role and presence of roadway-related factors

Tire's crash role (Vehicle classification)	Roadway related factors	Observed	Expected	Z-value
Active (Tire-related crash vehicle)	Adverse	72,005	32,736	6.9*
	None	124,963	164,232	-2.6*
Inactive (Other crash vehicle)	Adverse	573,661	612,930	-1.5*
	None	3,114,310	3,075,041	0.6

*Statistically significant at 95 percent confidence level
(Data Source: NMVCCS 2005-2007)

6.8 AGGRESSIVE DRIVING ACTS

Aggressive driving actions such as speeding, rapid or frequent lane changes, weaving, accelerating rapidly from stop, or stopping suddenly may expedite tire wear or tire failure.

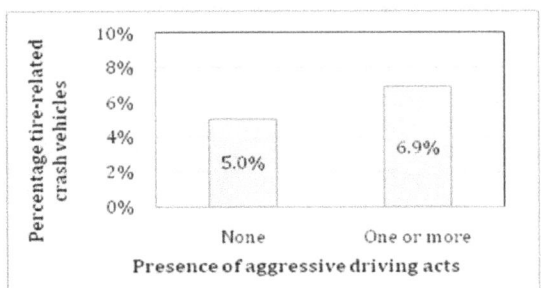

Figure 10. Percentage of tire-related crash vehicles in each category of aggressive driving acts' presence
(Data Source: NMVCCS 2005-2007)

Figure 10 shows that when drivers showed aggressive driving acts, 6.9 percent of the vehicles experienced tire problems. In the absence of such acts, a smaller percentage of vehicles were observed to experience tire problems (5.0%). The percentage of other crash vehicles for each category of aggressive driving acts' presence can be obtained by subtracting the percentage of tire-related crash vehicles from 100. For example, when drivers showed aggressive driving acts, about 93 percent were other crash vehicles while 6.9 percent of the vehicles experienced tire problems in the pre-crash phase.

In Table 11, there is no significant Z-value, which suggests that the differences between observed and expected frequencies for all cells are not statistically significant. That is, there is no sufficient evidence to reject the assumption that tire's crash role has no association with driver's aggressive driving behavior.

Table 11. Observed and expected frequencies by tire's crash role and presence of driver aggressive acts

Tire's crash role (Vehicle classification)	Presence of driver aggressive acts	Observed	Expected	Z-value
Active (Tire-related crash vehicle)	One or more	14,890	11,105	1.2
	None	173,733	177,518	-0.3
Inactive (Other crash vehicle)	One or more	200,600	204,385	-0.2
	None	3,270,876	3,267,091	0.1

*Statistically significant at 95 percent confidence level
(Data Source: NMVCCS 2005-2007)

6.9 CLIMATIC CONDITIONS

Under extreme temperatures, tires are vulnerable to tire degradation, significant loss of tire pressure, additional flexing, and stress on the sidewalls. These tire conditions may lead to tire failure or even blow out. In this analysis, three climatic conditions, cold (November to February), hot (July to September), and mild (March to June, October) are considered based on the month of the year in which a crash occurred. In the categorization of climatic conditions, temperature variations by State or location are not considered due to lack of information.

The NMVCCS data shows that 5.5 percent of the vehicles running in hot weather and 5.3 percent in cold weather experienced tire problems in the pre-crash phase. These percentages are higher than the percentage (4.6 %) under mild climatic conditions (Figure 11). The percentage of other crash vehicles for each category of climatic conditions can be obtained by subtracting the percentage of tire-related crash vehicles from 100. For example, in hot weather 94.5 percent were other crash vehicles while 5.5 percent were tire-related crash vehicles.

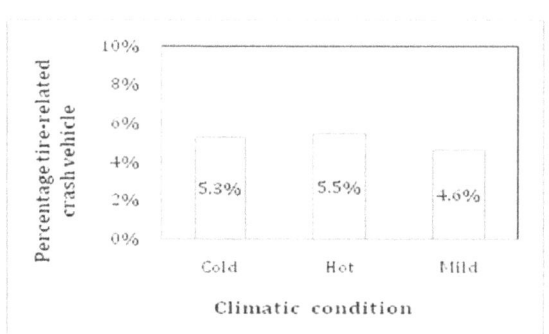

Figure 11. Percentage of tire-related crash vehicles by climatic condition
(Data Source: NMVCCS 2005-2007)

CFA results show how the climatic conditions affect tire's crash role. In Table 12, there is no significant Z-value, which suggests that the differences between observed and expected frequencies for all cells are not statistically significant. That is, there is no sufficient evidence to reject the assumption that tire's crash role has no association with climatic conditions. Note that in this analysis, the categorization of climatic conditions did not take into account temperature variations by State or location.

Table 12. Observed and expected frequencies by tire's crash role and climatic condition

Tire's crash role (Vehicle classification)	Climatic condition	Observed	Expected	Z-value
Active (Tire-related crash vehicle)	Cold	66,661	64,045	0.3
	Hot	63,004	58,588	0.5
	Mild	67,756	74,788	-0.9
Inactive (Other crash vehicle)	Cold	1,196,746	1,199,362	-0.1
	Hot	1,092,754	1,097,170	-0.1
	Mild	1,407,587	1,400,555	0.2

*Statistically significant at 95 percent confidence level
(Data Source: NMVCCS 2005-2007)

7. SUMMARY AND DISCUSSION

Tire problems are inherently hazardous to vehicle safety. When these problems emerge in the pre-crash phase, the time window for attempting a crash avoidance maneuver is normally very small. The NMVCCS records tire-related issues in the pre-crash phase as associated factors, critical pre-crash events, and critical reasons for the critical pre-crash event. This study is focused on these perspectives of tire-related crash vehicles.

About 5 percent of the estimated total number of vehicles involved in the NMVCCS crashes experienced tire problems during crash occurrence. Passenger cars accounted for about 66 percent of such vehicles. The findings from CFA highlighted some of the crash factors that have significant association with a vehicle experiencing tire problems in the pre-crash phase. With prior tire damage, a vehicle experiencing tire problems is significantly more likely than chance. This is also the case if a vehicle runs on tires overinflated by more than 25 percent of the recommended tire pressure. A crash vehicle may experience tire problems during crash occurrence if it runs on tires with inadequate tread depth. Rollover is also associated with vehicles that experienced tire problems in the pre-crash phase. The vehicles running under adverse roadway conditions such as wet roads may become more vulnerable to tire problems. Concerning driver factors, both the lack of driving experience and lack of familiarity with the vehicle are likely to contribute to a vehicle experiencing tire problems during crash occurrence.

Thus, while tire problems themselves increase the potential of a vehicle's involvement in crashes, other crash factors (an inexperienced driver, adverse roadway conditions, etc.) may add to the crash risk due to their influence on a vehicle experiencing these problems in the pre-crash phase. The findings of this study emphasize the importance of careful monitoring of tread depth as well as maintaining the proper inflation pressure of all tires of the vehicle. This monitoring and maintenance can also provide safeguards against the emergence of tire problems that are likely to appear under adverse road conditions. In addition, less experienced drivers or people not familiar with the vehicles they are driving should be more cautious to prevent tire-related crashes.

8. APPENDIX

Z-statistics based on first order CFA are obtained as follows:

$$Z_{ij} = \frac{O_{ij} - E_{ij}}{\sqrt{E_{ij}}} \frac{\sqrt{n-1}}{\sqrt{N}\sqrt{deff}},$$

where O_{ij} is the weighted observed frequency, E_{ij} is the weighted expected frequency, n is the sample size, N is the weighted total, and *deff* is the design effect. Here, *deff*, the design effect is the ratio of the variance of a statistic with a complex sample design used in NMVCCS to the variance of that statistic with a simple random sample. It is computed by using Taylor series expansion.

First order CFA assumes that the variables forming the contingency table under study
(1) may show main effect, and
(2) are totally independent of each other.

The observed weighted frequency for cell (i,…,l) is

$$O_{i,\ldots,l} = \sum_{k \in S} w_k y_{ki,\ldots,j},$$

where $y_{ki\ldots j} = \begin{cases} 1 & \text{if a observation unit k is in cell } (i,\ldots,l) \\ 0 & \text{other wise} \end{cases}$ and w_k is the weight for observation unit k

A maximum likelihood estimator for the expected cell frequency is

$$E_{i,\cdots,l} = \frac{O_{i\ldots} \times O_{.j..} \times \cdots \times O_{\ldots l}}{N^{(d-1)}},$$

where i^{th} of d variables has C_i categories with i=1,…, C_1, j=1,…, C_2, l=1,…, C_d and N is the weighted total.

<u>When only two variables (d=2) are considered,</u>

The observed weighted frequency for cell (i,j) is

$$O_{ij} = \sum_{k \in S} w_k y_{kij},$$

Where $y_{kij} = \begin{cases} 1 & \text{if a observation unit k is in cell } (i,j) \\ 0 & \text{other wise} \end{cases}$ and w_k is the weight for observation unit k

The weighted expected frequency is

$$E_{ij} = \frac{O_{i.} \times O_{.j}}{N^{(2-1)}},$$

Where $O_{i.}$ is the weighted row sum and $O_{.j}$ is the weighted column sum and N is $O_{..}$, the weighted total sum. More details are provided in Von Eye[8] and Lohr.[10]

9. REFERENCES

[1] NHTSA. (2003, June). "FMVSS No. 139, New Pneumatic Tires for Light Vehicles", Docket No. NHTSA-2003-15400-2.

[2] *Tire Pressure Monitoring Systems*, 49 C.F.R. § 571.138 (2011).

[3] *New Pneumatic Radial Tires for Light Vehicles*, 49 C.F.R. § 571.139 (2011).

[4] NHTSA. (2008, July). National Motor Vehicle Crash Causation Survey: Report to congress. (Report No. DOT HS 811 059). Washington, DC: National Highway Traffic Safety Administration. Available at www-nrd.nhtsa.dot.gov/Pubs/811059.pdf

[5] Institute for Research in Public Safety [now known as the Transportation Research Center at Indiana University]. (1979, May). Tri-Level Study of the Causes of Traffic Accidents. Bloomington, IN: Author.

[6] NHTSA (2008, December). National Motor Vehicle Crash Causation Survey: Field Coding Manual. (Report No. DOT HS 811 051). Washington, DC: National Highway Traffic Safety Administration. Available at www-nrd.nhtsa.dot.gov/Pubs/811051.pdf.

[7] Von Eye, A. (1990), Introduction to Configural Frequency Analysis. New York: Cambridge University Press.

[8] Von Eye, A. (2002), Configural Frequency Analysis, Hillsdale, NJ: Lawrence Erlbaum.

[9] SAS/ETS(R) 9.1.3 User's Guide. (1999). Cary, NC: SAS Institute Inc.

[10] Lohr, S. L. (1999), Sampling: Design and Analysis, Pacific Grove, CA: Duxbury Press.

DOT HS 811 617
April 2012

U.S. Department of Transportation
National Highway Traffic Safety Administration

8603-042412-v2

www.ingramcontent.com/pod-product-compliance
Lightning Source LLC
Chambersburg PA
CBHW081819170526
45167CB00008B/3459